# Exploring Measurement

Length • Area • Volume • Mass • Time

Grades 2-3

Rik Carter

World Teachers Press

Published with the permission of R.I.C. Publications Pty. Ltd.

Copyright © 1996 by Didax, Inc., Rowley, MA 01969. All rights reserved.

First published by R.I.C. Publications Pty. Ltd., Perth, Western Australia.

Limited reproduction permission: The publisher grants permission to individual teachers who have purchased this book to reproduce the blackline masters as needed for use with their own students. Reproduction for an entire school or school district or for commercial use is prohibited.

Printed in the United States of America.

Order Number 2-5007
ISBN 1-885111-19-3

A B C D E F 96 97 98 99

*Educational Resources*

395 Main Street
Rowley, MA 01969

# Foreword

Grades 2-3

The *Exploring Measurement* series provides a rich source of measurement activities for elementary and middle school students. The activities cover the five areas of measurement, which are length, area, volume and capacity, mass and time. An objective is provided for each activity.

## Contents

| Contents | Pages | Objectives |
|---|---|---|
| Teacher Information | 4-5 | |
| **Length** | | |
| Measuring Animals 1 | 6 | Use arbitrary units to measure heights |
| Measuring Animals 2 | 7 | Use arbitrary units to measure and order lengths |
| Handspans | 8 | Estimate length before measuring |
| Measuring with a Pencil | 9 | Use arbitrary units to measure and order length, estimate first |
| Bricks and Walls | 10 | Arrange shapes to cover a larger area |
| Find the Difference | 11 | Use standard units to measure length |
| Measuring with Squares | 12 | Use standard units to measure length and width |
| Snake Measure | 13 | Use a nonstandard consistent unit to measure length, height and width |
| Measuring Body Parts | 14 | Use standard units to measure body parts |
| Walking Time | 15 | Relate the measure of length to the measure of time |
| **Area** | | |
| Coloring Triangles | 16 | Find area of a triangle using an arbitrary unit |
| Shape Coloring | 17 | Order the area of various shapes by the time it takes to color them |
| Covering with Counters | 18 | Use counters to cover areas of shapes, estimate first |
| Shapes | 19 | Match shapes by area and configuration |
| Making Patterns | 20 | Arrange octagons into patterns |
| Ordering | 21 | Order the area of shapes by measuring with cubes or counters |
| Cover, Make and Fill | 22 | Relate the measure of area to the measure of capacity |
| Triangles or Squares? | 23 | Use triangles and squares as arbitrary units for measuring area |
| Cover it! | 24 | Compare the areas of classroom objects |
| Cut it Out! | 25 | Compare the area of shapes using an arbitrary unit |
| **Volume and Capacity** | | |
| Measuring with Rice | 26 | Use rice to measure the capacities of containers, estimate first |
| How Many Marbles? | 27 | Compare and order the capacity of different containers |
| Apricot Crunchies | 28 | Discuss volume and capacity through cooking |
| Handful of Sand | 29 | Compare and graph the capacity of containers using sand |
| Measuring with Macaroni | 30 | Relate the capacity of containers to their circumference |
| Measuring with Cubes | 31 | Measure the capacity of containers using arbitrary units |
| Measuring with Water | 32 | Measure and order containers by their capacity using water |
| Cube Layers | 33 | Use cubes to measure capacity |
| Rising Water | 34 | Use water displacement to order and measure volume |
| Cube Power | 35 | Discover patterns in cubic volume |
| **Mass** | | |
| Making and Comparing | 36 | Compare the mass of constructed models |
| Equal Weight | 37 | Directly compare the mass of objects by balancing |
| Weighing with Cubes | 38 | Compare and order objects according to mass |
| Covering and Weighing | 39 | Relate and compare the measure of area to mass |
| Your Weight | 40 | Measure personal mass and compare this to other masses |
| Making it Balance | 41 | Use a balance scale to balance various objects |
| Ordering by Weight | 42 | Order objects by mass through estimation and check results |
| Measuring in Ounces | 43 | Measure the mass of objects in ounces |
| Measuring Mass | 44 | Measure the mass of students in pounds |
| Length and Mass | 45 | Relate the measure of mass to the measure of length |
| **Time** | | |
| Time | 46 | Discuss times of the day, week and year |
| Days and Months | 47 | Order the days of the week and months of the year |
| Timing | 48 | Measure the duration of time using an arbitrary unit |
| Just a Minute! | 49 | Complete activities to experience one minute |
| Alphabet Hunt | 50 | Estimate the time of events |
| One and Two Minutes | 51 | Develop an awareness of one and two minutes |
| June Quiz | 52 | Read and comprehend a monthly calendar |
| Weekly Events | 53 | Complete a weekly diary |
| Clock Reading 1 | 54 | Read time on the hour |
| Clock Reading 2 | 55 | Read time on the half-hour, quarter-hour and five minutes |

# Teacher Information

## Introduction

Measurement is a vital part of any mathematics curriculum as it has an integral role in nearly all forms of math and also plays a major influence in our daily lives. As we wake in the morning we are influenced by the measurement of time, we see measurement in most of the food items we have for breakfast and indeed measure our food intake as part of our diet. There are very few parts of our daily lives that are not influenced by measurement in some form.

It is very important to make students aware of the different forms of measurement. Often an overemphasis can be placed on the measurement of length which provides an unbalanced concept with students. Below are outlined the main forms of measurement, all of which are extensively covered in this series of books.

The ability of students to understand the concepts of measurement is developmental. For example, the concept of length is understood before the concept of area and so on. Measurement is best tackled from the area of spatial awareness as in most cases we are talking about the measurement of our daily surroundings or events.

**Arbitrary Units:** the use of units of measure that students can relate to are a vital aspect in the introductory stages. Using hand spans, desk widths, fingers, rice and many more simple units will assist with early understanding.

## Length

The measurement of length is an area that becomes very obvious to students. One reason for this is that one of the first tools of learning that a student uses is a ruler. Another is the interest student have in their own growth which is a measurement of length.

In the measurement of length it is important to emphasis accuracy and to develop this accuracy through increasingly small units of measure (feet, inches, parts of an inch, etc.). Students should not move to smaller units of measure until they are confident with the ones they are working with and identify the need for smaller units.

## Area

The measurement of area can be a difficult concept to grasp. Use of concrete materials and arbitrary units of measure are recommended to ensure a good understanding of this concept, before specific units of measure and formula are introduced.

## Volume and Capacity

The measurement of volume and capacity is also an area that requires a good grounding of concrete activities and the use of arbitrary units. It can be a difficult concept to understand and a lot of introductory activity is suggested.

## Time

Once students are taught how to tell the time, this area of measurement is often ignored. However there are many areas within the concept of time that will prove fascinating to students, as well as being very important (calendars, diaries, timelines, time zones, etc).

## Mass

As there are specific apparatus that are used for the measurement of mass (scales of various forms) this form of measurement can be developed at an early stage, using common objects including the students themselves. "Mass" is an important term and should be encouraged ahead of the use of the word "weight".

# Teacher Information - Example Lesson Development

The following is a lesson development using one of the pages in this book. It is an example of how the activity could be introduced, developed and extended.

**Activity** — Measuring Animals 1: page 6

## Introductory Work

When introducing measurement as a concept to students it is very important that they are able to relate measurement to their own lives and their immediate surroundings. It is also important to make students aware that measurement occurs in different forms (linear, area, time, etc.) The use of arbitrary units is very important. This allows students to use easily identifiable units to measure given objects. Identifying the arbitrary units that are available in the classroom is a good introduction activity which will promote discussion and interest in the activities to follow.

## Completing the Worksheets

The following is a suggestion for the development and extension of this activity.

1. Discuss why blocks are an easy measurement tool. Then talk about which size block would be best for this activity. Do we have to use the same size block for each animal? etc.

2. Have students round measurements off to even blocks, to avoid the use of fractions. Discuss how this could produce some variety in the answers.

3. Have students check their measurements twice for accuracy.

4. The concept of tallest and shortest can be developed into other units of measurement (longest, shortest, etc.)

*Note:* Be sure students are aware that they are measuring drawings of animals and discuss what they might need to measure the real animal.

## Extension

The extension of the activity is largely covered by the next activities, however further discussion and use of arbitrary units is important and using blocks to measure other objects in the classroom is a useful extension.

# Measuring Animals 1

Name _____

Use blocks to measure the height of each of the animals.

Which animal is the tallest?

_____

Which animal is the shortest?

_____

# Measuring Animals 2

Name _____

Use blocks to measure the length of each of the animals.

Which is the longest animal? _____

Which animal is the shortest? _____

Order the animals from longest to shortest. _____

_____

Exploring Measurement, Grades 2–3       World Teachers Press       7

# Hand Spans

Name _____

Use your hand span to measure each of the objects listed below. Make a guess before you measure.

1. The length of your desk.

   Estimate _____ Measure _____

2. The length of your arm.

   Estimate _____ Measure _____

3. The length of your leg.

   Estimate _____ Measure _____

4. The height of a door.

   Estimate _____ Measure _____

5. The length of your shoe.

   Estimate _____ Measure _____

6. The length of the chalkboard.

   Estimate _____ Measure _____

Order the objects you have measured from longest in length to shortest in length.

1. _____   4. _____

2. _____   5. _____

3. _____   6. _____

# Measuring with a Pencil  Name _____

Use a pencil to measure each of the objects listed below. Make an estimate before you measure.

| Object | Estimate | Measure |
|---|---|---|
| A book | | |
| A chair | | |
| A desk | | |
| A ruler | | |
| A sheet of paper | | |
| Your arm | | |
| Your leg | | |

Order the objects you have measured from longest in length to shortest in length.

_____     _____

_____     _____

_____     _____

_____     _____

# Bricks and Walls

Name _____

Color and cut out the bricks at the bottom of the page. Can you put them on the blank wall below so they fit exactly?
There may be more than one answer.

How many answers did you find? _____

# Find the Difference

Name _____

Place half inch cubes on the rectangles below to find the difference in length.

1.  [ ]          [ ]          Difference _____

2.  [ ]          [ ]          Difference _____

3.  [ ]          [ ]          Difference _____

4.  [ ]          [ ]          Difference _____

5.  [ ]          [ ]          Difference _____

6.  [ ]          [ ]          Difference _____

7.  [ ]   [ ]                 Difference _____

Exploring Measurement, Grades 2–3     World Teachers Press

# Measuring with Squares  Name _____

Look carefully at the drawings in the grid below.

What is the length and width in squares of each of the drawings?

The butterfly is  _____ long  and  _____ wide.

The bicycle is  _____ long  and  _____ wide.

The horse is  _____ long  and  _____ wide.

The lion is  _____ long  and  _____ wide.

The pencil is  _____ long  and  _____ wide.

# Snake Measure

Name _____

Cut out the measuring snake at the bottom of this page and glue it to a piece of cardboard. Using this snake, measure the lengths of the classroom objects listed in the table and record your results.

| Object | Snake Measures |
|---|---|
| The length of your desk | |
| The height of your desk | |
| The length of a book | |
| The width of your chair | |
| The height of your chair | |
| The height of the door | |
| The width of the door | |
| The depth of the door | |
| The length of a pencil | |
| Own choice | |
| Own choice | |

Exploring Measurement, Grades 2-3

# Measuring Body Parts

Name _____

With a partner's help, measure the body parts listed below. Use a tape measure.

My head is _____ inches around.

I am _____ inches tall.

My neck is _____ inches around.

My thumb is _____ inches long.

My waist is _____ inches around.

My wrist is _____ inches around.

My hand span is _____ inches.

My knee is _____ inches around.

My foot is _____ inches long.

14    World Teachers Press    Exploring Measurement, Grades 2-3

# Walking Time

Name _____

First estimate and then measure the time taken to walk fifty steps.

Estimate _____     Time _____

Do the same for these distances.

| Distances (steps) | Estimated time | Actual time |
|---|---|---|
| Five steps | | |
| Ten steps | | |
| Thirty steps | | |
| One hundred steps | | |
| Two hundred steps | | |

Is the length of your step always the same? Why?

_____

_____

_____

Is a step a good way of measuring distances? Why?

_____

_____

_____

Exploring Measurement, Grades 2-3    World Teachers Press

# Coloring Triangles

Name _____

Color the top four small triangles red.
Color the bottom five small triangles blue.
Cut out all the small triangles and place them exactly over one of the large triangles below.
Make a pattern. Make other patterns on the last three large triangles.

16     World Teachers Press     Exploring Measurement, Grades 2–3

# Shape Coloring

Name _____

If you had to color the shapes below, how long do you think each shape would take you?

Shape 1 _____

Shape 2 _____

Shape 3 _____

Shape 4 _____

Color each shape.

Order the shapes from longest time to color to least time to color.

_____

Order the shapes from largest to smallest.

_____

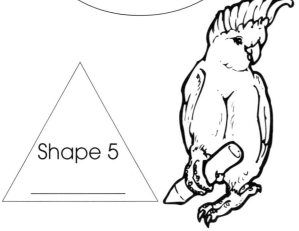

Shape 5 _____

Are the orders the same? _____

# Covering with Counters

Name _____

Cover the shapes below with counters. Estimate the number of counters you will need for each shape before you start.

Estimate ____

Actual ____

Estimate ____

Actual ____

Estimate ____

Actual ____

Estimate ____

Actual ____

Estimate ____

Actual ____

# Shapes

Name _____

Cut out the shapes at the bottom of the page and glue them on the picture that matches each shape.

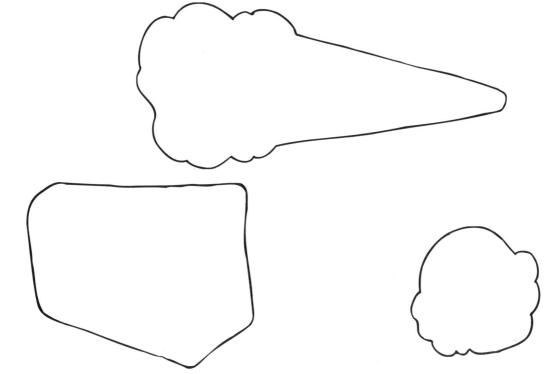

Exploring Measurement, Grades 2–3

# Making Patterns

Name _____

Color the octagonal shapes below.
Cut them out. Arrange them to make an interesting pattern. Glue this pattern on a piece of paper.

# Ordering

Name _____

Cover the rabbits below with cubes or counters.
Number the rabbits in order from biggest to smallest.

Exploring Measurement, Grades 2–3   World Teachers Press

# Cover, Make and Fill

Name _____

Use cubes to cover this shape.

How many did you use? _____

Cut along the lines of this shape. Fold along the dotted lines of this shape. Use glue to make this box. Fill your box with the cubes.

How many did you use? _____

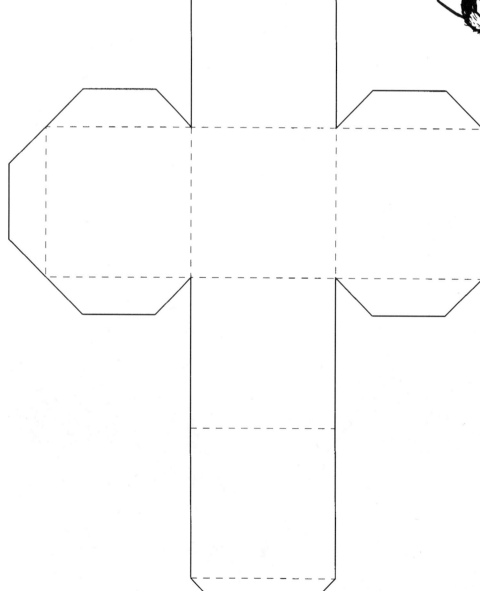

22     World Teachers Press     Exploring Measurement, Grades 2–3

# Triangles or Squares?

Name _____

Which of these two shapes, the triangle or the square, will cover the large rectangle below using the least number of shapes? Cut out the triangle and the square and trace around them in the large rectangle.
Use a different color for each.

Number of squares _____       Number of triangles _____

Exploring Measurement, Grades 2-3

# Cover It!

Name _____

How many copies of this worksheet do you think you would need to cover the objects below?

| Object | Estimate | Number |
|---|---|---|
| Your desk | | |
| Classroom window | | |
| Classroom door | | |
| Your chair | | |
| A chalkboard | | |
| A television screen | | |
| Own choice | | |
| Own choice | | |

What problems did you have measuring the area of the above objects?

_____

_____

_____

What else could you use to measure area?

_____

# Cut it Out!

Name _____

Cut out the small square and use it to measure the area of the other shapes.
Estimate the number you will need before you measure.

Estimate _____

Measure _____

Estimate _____

Measure _____

Estimate _____

Measure _____

Estimate _____   Measure _____

Exploring Measurement, Grades 2–3      World Teachers Press      25

# Measuring with Rice

Name _____

Using containers like the ones shown below, estimate how many spoonfuls of rice will fill each container. Then measure and write the number.

Estimate _____

Measure _____

Estimate _____

Measure _____

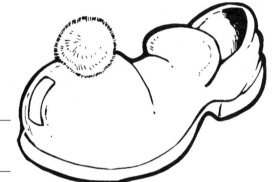

Estimate _____

Measure _____

Estimate _____

Measure _____

Estimate _____

Measure _____

# How Many Marbles?

Name _____

How many marbles will fill each of the containers below? Make an estimate before you measure.

|  | Estimate | Measure |
|---|---|---|
| Your hand | _____ | _____ |
| Two hands | _____ | _____ |
| A pencil case | _____ | _____ |
| A cup | _____ | _____ |

Order your containers from largest to smallest

_____

How many marbles will fill each of the containers below? Make an estimate before you measure.

|  | Estimate | Measure |
|---|---|---|
| Your hand | _____ | _____ |
| Two hands | _____ | _____ |
| A pencil case | _____ | _____ |
| A cup | _____ | _____ |

Order your containers from largest to smallest

_____

Are the two orders the same? Why? _____

Exploring Measurement, Grades 2–3     World Teachers Press     27

# Apricot Crunchies

Name _____

**You will need:**
a baking tray
a medium-size mixing bowl
a knife and dessert spoon
a chopping board and teaspoon
a wire cooling rack

**Ingredients:**
1 can of condensed milk
9 oz. of chopped dried apricots
9 oz. rolled oats
1 teaspoon of cinnamon
margarine

**Method**
1. Lightly grease the baking tray with a little margarine.
2. Mix the milk, oats and cinnamon in the bowl.
3. Add the apricots and stir thoroughly with the spoon.
4. Put spoonfuls of the mixture on the baking tray about an inch apart.
5. Bake in a moderate oven for about 15 minutes or until brown.
6. When cooked, place on a wire rack to cool.

How many crunchies did your recipe make? _____

Write a sentence or two about making apricot crunchies.

_____

_____

_____

_____

# Handful of Sand

Name _____

How many handfuls of sand will fill each of the containers listed beneath the graph? Record your answer on the graph.

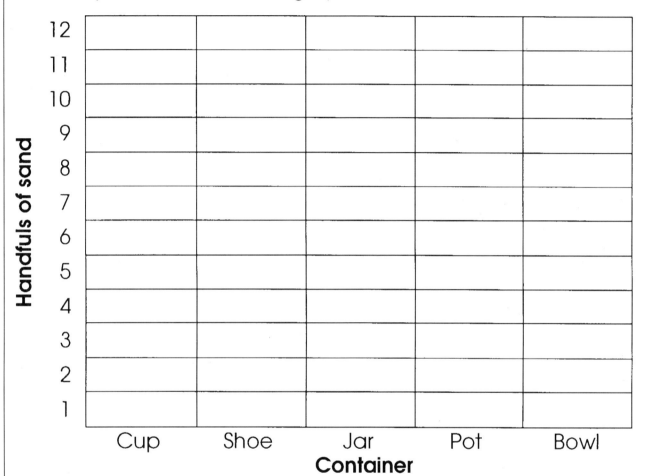

Will everyone in your class get the same answer? Why?

_____

Which container held the most? _____

Which container held the least? _____

What other things besides sand could be used to measure how much a container could hold?

_____

Exploring Measurement, Grades 2–3    World Teachers Press    29

# Measuring with Macaroni

Name _____

How many spoonfuls of macaroni will fill each of these containers?

| | | |
|---|---|---|
| A coffee mug | _____ spoons | ☐ |
| A small can | _____ spoons | ☐ |
| Your hand | _____ spoons | ☐ |
| A drinking glass | _____ spoons | ☐ |

Number your containers 1, 2, 3, and 4, starting with the one which held the most and ordering downwards.

Use string to measure the distance around each container. Glue the strings on a piece of paper from longest to shortest.

When you have completed your graph, answer the questions below.

Did the container with the longest string hold the most macaroni?

Why? _____

_____

Did the container with the shortest string hold the least macaroni? Why

_____

_____

# Measuring with Cubes    Name _____

Use cubes to fill the containers listed below. Make an estimate first of the number of cubes needed.

| Container | Estimate | Measure |
|---|---|---|
| Margarine container | | |
| Glass jar | | |
| Milk carton | | |
| Yogurt container | | |
| Coffee mug | | |
| Your hand | | |
| Own choice | | |
| Own choice | | |

What problems did you have when filling the containers?

_____

_____

Can you find a container that will hold four times the amount of cubes as the margarine container?

_____

_____

Exploring Measurement, Grades 2–3        World Teachers Press

# Measuring with Water

Name _____

Use water to fill the containers listed below. Make an estimate first of the order of the containers from least to most. Then fill with water to find the actual order.

|  | Estimate | Actual |
|---|---|---|
| 1. Milk carton | ☐ | ☐ |
| 2. Glass jar | ☐ | ☐ |
| 3. Drink can | ☐ | ☐ |
| 4. Milk bottle | ☐ | ☐ |
| 5. Drink bottle | ☐ | ☐ |
| 6. Small bucket | ☐ | ☐ |
| 7. Pot | ☐ | ☐ |
| 8. Bowl | ☐ | ☐ |

Were you surprised by the results? Why?

_____
_____
_____
_____

# Cube Layers

Name _____

Use cubes to build the object drawn here.

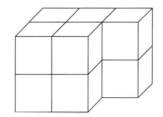

Now add a second layer.

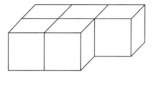

Keep adding layers to the object and record the numbers of cubes used in the table below. Make an estimate first.

| Number of layers | Estimate of cubes used | Actual cubes used |
|---|---|---|
| Two | | |
| Three | | |
| Four | | |
| Five | | |
| Six | | |
| Seven | | |
| Eight | | |
| Nine | | |
| Ten | | |

How many cubes would you need for twenty layers?

_____

# Rising Water

Name _____

Fill a jar half full with water. Place each of the objects listed below in the jar, one at a time. Record the increase in the water level on the side of the jar with a felt pen. Place the objects in order from largest to smallest, but estimate their order first.

Estimate    Actual

1. Ball of modeling clay
2. Ten marbles
3. A golf ball
4. A rock
5. A lemon
6. An apple
7. A baseball
8. A potato
9. Own choice
10. Own choice

Did you have any problems with the measuring?
If so, tell what happened.

_____
_____

# Cube Power

Name _____

Here is a cube of one cube.

Here is the next cube. Build it.
How many cubes did you use?

_____

Here is the next cube. Build it.
How many cubes did you use?

_____

Draw the next cube, then build it.
How many cubes did you use?

_____

Look for a pattern in the building
of these cubes. Once you have
found the pattern, explain it.

_____

_____

_____

Exploring Measurement, Grades 2-3    World Teachers Press    35

# Making and Comparing

Name _____

Use modeling clay to make the animals drawn below.

Order your animals from heaviest to lightest.
Make an estimate first.

 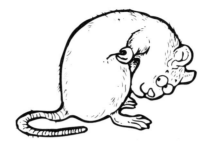 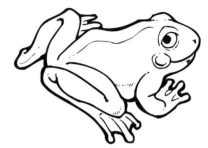

  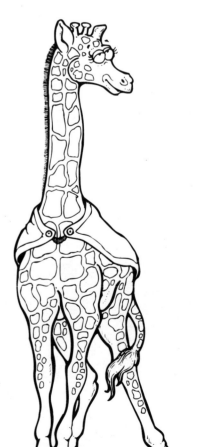

|  | Estimate | Actual |
|---|---|---|
| giraffe |  |  |
| frog |  |  |
| snail |  |  |
| horse |  |  |
| dog |  |  |
| mouse |  |  |

# Equal Weight

Name _____

Find the objects drawn below, then find other objects that weigh the same. Draw them.

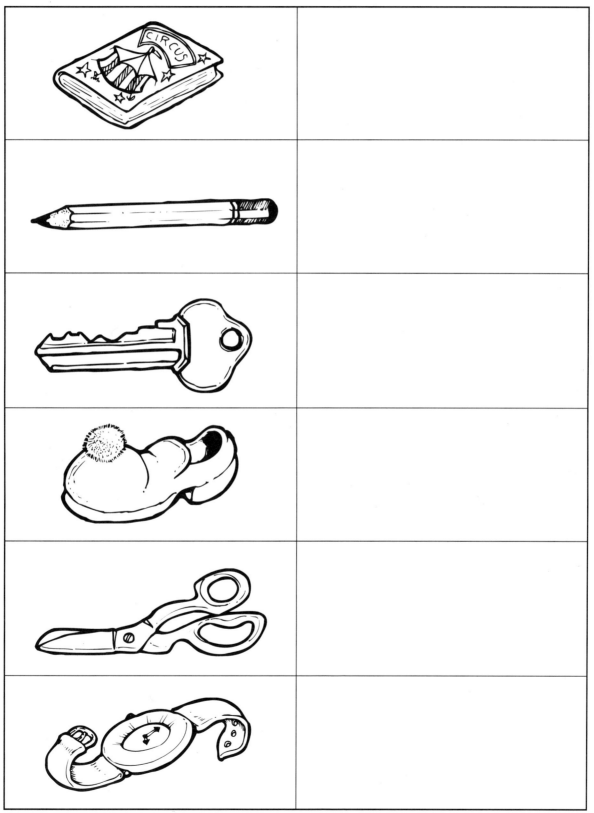

Exploring Measurement, Grades 2–3     World Teachers Press

# Weighing with Cubes

Name _____

Use a balance scale to weigh the objects shown below, using cubes to balance the objects. Record the number of cubes used.

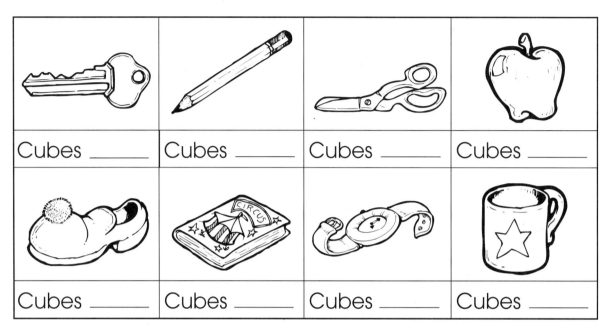

| Cubes ___ | Cubes ___ | Cubes ___ | Cubes ___ |
| Cubes ___ | Cubes ___ | Cubes ___ | Cubes ___ |

Cut out the pictures below.
Glue them on a piece of paper in order from lightest to heaviest.
Which two objects were closest in weight?

_____   _____

Which two objects were furthest apart in weight?

_____   _____

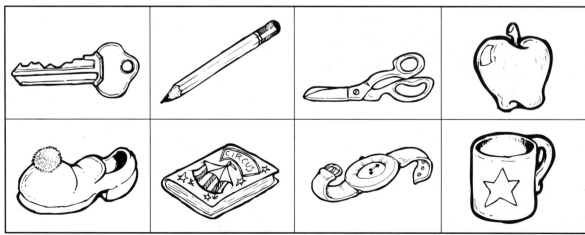

# Covering and Weighing

Name _____

You will need cubes and the objects shown below. You need to cover each of the objects with cubes and count the cubes you used. You will then need to use a balance scale to weigh the objects using cubes and count the cubes you used.

| Objects | Number for covering | Number for weighing |
|---|---|---|
| scissors | | |
| circus book | | |
| key | | |
| leaf | | |

Which objects used more cubes for covering? _____

_____

Which objects used more cubes for weighing? _____

_____

Explain your findings. _____

# Your Weight

Name _____

You will need a bathroom scale.
Weigh yourself. How much do you weigh? _____

List two objects that are heavier than you. How could you weigh these objects?

    Object 1 _____    Weight _____

    Object 2 _____    Weight _____

List two objects that are lighter than you. How could you weigh these objects?

    Object 1 _____    Weight _____

    Object 2 _____    Weight _____

Try to find two objects that are close to your weight. How could you weigh these objects?

    Object 1 _____    Weight _____

    Object 2 _____    Weight _____

Estimate which was the heaviest object? _____

Estimate which was the lightest object? _____

What problems did you have in completing this activity?

_____

_____

# Making it Balance

Name _____

What will balance the objects listed below? Use a balance scale.

| Object | How I made it balance |
|---|---|
| A pencil case | |
| 25 wooden cubes | |
| A book | |
| A shoe | |
| An apple | |
| A cup | |
| A rock | |
| A pencil | |
| A box of tissues | |
| A golf ball | |

Place the objects in order from lightest to heaviest.

_____

_____

Which objects were difficult to weigh. Why?

_____

_____

# Ordering by Weight

Name _____

Estimate how much the objects listed below weigh in order from lightest to heaviest.

| A book | A pencil | A shoe | An apple |
| An eraser | A sharpener | A rock | An orange |
| A golf ball | A marble | A potato | A pen |

1. _____  7. _____
2. _____  8. _____
3. _____  9. _____
4. _____  10. _____
5. _____  11. _____
6. _____  12. _____

Use a balance scale to check your estimates. Has it changed the order? Write the correct order below.

1. _____  7. _____
2. _____  8. _____
3. _____  9. _____
4. _____  10. _____
5. _____  11. _____
6. _____  12. _____

# Measuring in Ounces

Name _____

Measure the objects listed below with a balance scale.
Use proper weights.
Estimate the mass of each object before you measure.

| Object | Estimate | Weight in ounces |
|---|---|---|
| A pen | | |
| A pencil | | |
| An apple | | |
| A marble | | |
| An eraser | | |
| A shoe | | |
| An orange | | |
| A cup | | |
| 10 cubes | | |
| Own choice | | |

Put the objects in order from lightest to heaviest.

1. _____
2. _____
3. _____
4. _____
5. _____

6. _____
7. _____
8. _____
9. _____
10. _____

Exploring Measurement, Grades 2–3

# Measuring Mass

Name _____

Use a bathroom scale to measure your own mass and the mass of nine other students in your class.
Round off your measurements to the nearest pound.
Estimate the mass of each student first.

| Student | Estimated mass | Actual mass |
|---|---|---|
| Myself | | |
| | | |
| | | |
| | | |
| | | |
| | | |
| | | |
| | | |
| | | |
| | | |

Order the students you measured from lightest to heaviest.

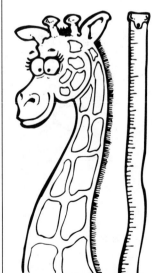

1. _____  6. _____

2. _____  7. _____

3. _____  8. _____

4. _____  9. _____

5. _____  10. _____

Is the tallest student the heaviest? _____

# Length and Mass

Name _____

Measure the length and weight of each object below.

| Object | Length | Mass |
|---|---|---|
| A pencil | | |
| A pair of scissors | | |
| A book | | |
| A piece of paper | | |
| A key | | |
| A shoe | | |

Put the objects in order from longest to shortest.

1. _____  4. _____

2. _____  5. _____

3. _____  6. _____

Put the objects in order from heaviest to lightest.

1. _____  4. _____

2. _____  5. _____

3. _____  6. _____

Is the longest object the heaviest? Why? _____

_____

Is the shortest object the lightest? Why? _____

# Time

Name _____

What time do you have lunch? _____
_____

On which day of the week do you have art class? _____

Which is your favorite day of the week? Why? _____
_____

Draw a picture of the things you like doing in Summer, Fall, Winter and Spring.

| Summer | Fall |
|---|---|
| Winter | Spring |

In which month is your birthday? _____

In which season is this? _____

On which day of the week is your birthday this year?
_____

# Days and Months

Name _____

Use a calendar for these activities.
Write the months of the year in order.

_____ _____ _____

_____ _____ _____

_____ _____ _____

_____ _____ _____

Write the days of the week in order.

_____ _____ _____

_____ _____ _____

_____

Which months are in:

Summer? _____

Fall? _____

Winter? _____

Spring? _____

Which month(s) have:

28 days? _____

30 days? _____

31 days? _____

# Timing

Name _____

With a friend, record the time it takes to complete the activities listed below.
Measure the time it takes by using hand claps.
Try to clap at an even rate.

| Activity | Your time | Friend's time |
|---|---|---|
| Say the alphabet | | |
| Say the 2 x table | | |
| Write the numbers 30 to 50 | | |
| Count to 100 | | |
| Count by 2s to 100 | | |
| Count by 10s to 200 | | |
| Bounce a tennis ball 20 times | | |
| Count from 100 down to 50 | | |
| Own choice | | |

Explain any problems you found.

_____

_____

# Just a Minute!

Name _____

Your teacher will time you for exactly one minute.
How many times can you do the activities listed below in one minute?

Count to 10.

Say the alphabet.

Print your name.

Count by 2s to 50.

Write your address.

Count backwards from 20 to 0.

Write your telephone number.

Blow up and deflate a balloon.

Write the alphabet.

Write the alphabet backwards.

Write the numbers 1 to 10.

Count from 20 to 40.

Read this worksheet.

# Alphabet Hunt

Name _____

Find and circle each letter of the alphabet in the box below.
Do this in alphabetical order.

How long do you think it will take you? _____
Have someone time you.

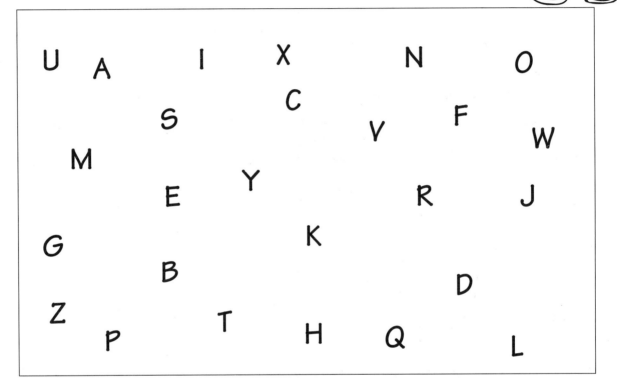

How long did it really take you to complete the activity?

_____

Repeat the activity.
Do you think it will take a shorter time? Why?

_____

How long did it take you on your second try?

_____

# One and Two Minutes

Name _____

Estimate the number of times you can do the activities listed below in one minute and then in two minutes.
Then do the activities to see how close your estimates were.
You will need someone to time you.

| Activity | 1 minute | | 2 minutes | |
| --- | --- | --- | --- | --- |
| | Est | Actual | Est | Actual |
| Bounce a ball | | | | |
| Print your name backwards | | | | |
| Say 'Humpty Dumpty' | | | | |
| Count to twenty | | | | |
| Build a tower with ten cubes | | | | |
| Count by sevens to seventy | | | | |

What interesting things did you notice about the results?

_____

_____

# June Quiz

Name _____

Below is a calendar for the month of June. Using the calendar answer the questions below.

## June

| Sun | Mon | Tues | Wed | Thurs | Fri | Sat |
|-----|-----|------|-----|-------|-----|-----|
|     |     |      | 1   | 2     | 3   | 4   |
| 5   | 6   | 7    | 8   | 9     | 10  | 11  |
| 12  | 13  | 14   | 15  | 16    | 17  | 18  |
| 19  | 20  | 21   | 22  | 23    | 24  | 25  |
| 26  | 27  | 28   | 29  | 30    |     |     |

What is the first day of the month? _____

What is the last day of the month? _____

How many Tuesdays are there in the month? _____

How many Sundays are there in the month? _____

What will be the first day of July? _____

What was the last day of May? _____

What season is June in? _____

52     World Teachers Press     Exploring Measurement, Grades 2-3

# Weekly Events

Name _____

Write a sentence and draw a picture of a special event that happened on each day of the week.

|  | | |
|---|---|---|
| Sun | | |
| Mon | | |
| Tues | | |
| Wed | | |
| Thur | | |
| Fri | | |
| Sat | | |

# Clock Reading 1

Name _____

Read the time on each clock below. Write the time next to the clock. One has been done for you.

 6:00

54     World Teachers Press     Exploring Measurement, Grades 2-3

# Clock Reading 2

Name _____

Read the time on each clock. Write the time next to the clock. One has been done for you.

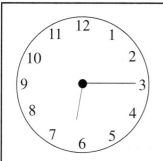  6:15

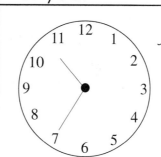

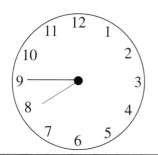

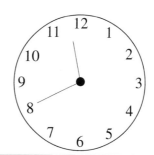

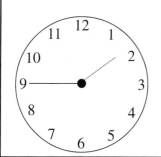

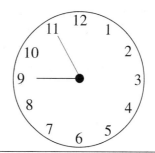

# About the Author

**Rik Carter**, Perth, Western Australia.
Rik has over 15 years experience in Australian primary schools as a classroom teacher, schools advisory teacher and administrator. He has a special interest in mathematics and has used his classroom and advisory experience to produce a range of titles for primary schools.